Inside the Military

MILITARY VEHICLES

Emma Bassier

DiscoverRoo
An Imprint of Pop!
popbooksonline.com

abdobooks.com

Published by Pop!, a division of ABDO, PO Box 398166, Minneapolis, Minnesota 55439.

Printed in China.

052019
092019

THIS BOOK CONTAINS RECYCLED MATERIALS

Cover Photo: US Department of Defense
Interior Photos: US Department of Defense, 1; Defense Visual Information Distribution Service, 5, 8–9, 19, 20, 21, 22, 23, 26–27, 28, 29, 30; iStockphoto, 6; Shutterstock Images, 7; National Motor Museum/Heritage Image Partnership Ltd/Alamy, 11; Paul Briden/Alamy, 12; US National Archives and Records Administration, 13, 15, 17 (top), 31; Chronicle/Alamy, 14; Scherl/Süddeutsche Zeitung Photo/Alamy, 16 (top); Library of Congress, 16 (bottom); amer ghazzal/Alamy, 17 (bottom), 25

Editor: Connor Stratton
Series Designer: Jake Slavik

Library of Congress Control Number: 2018964860

Publisher's Cataloging-in-Publication Data

Names: Bassier, Emma, author.

Title: Military vehicles / by Emma Bassier.

Description: Minneapolis, Minnesota : Pop!, 2020 | Series: Inside the military | Includes online resources and index.

Identifiers: ISBN 9781532163883 (lib. bdg.) | ISBN 9781644940617 (pbk.) | ISBN 9781532165320 (ebook)

Subjects: LCSH: Military vehicles--Juvenile literature. | Combat vehicles--Juvenile literature. | Armored military vehicles--Juvenile literature.

Classification: DDC 623.747--dc23

WELCOME TO DiscoverRoo!

Pop open this book and you'll find QR codes loaded with information, so you can learn even more!

Scan this code* and others like it while you read, or visit the website below to make this book pop!

popbooksonline.com/military-vehicles

*Scanning QR codes requires a web-enabled smart device with a QR code reader app and a camera.

TABLE OF CONTENTS

CHAPTER 1

MILITARY VEHICLES IN ACTION

A soldier drives a tank into battle. Another soldier controls the tank's main gun. He looks at a screen to aim. Then, a bullet from an enemy gun hits the tank. But the bullet bounces off the tank.

WATCH A VIDEO HERE!

The tank has thick **armor**. This hard layer protects the soldiers inside.

One common US tank fires rounds at speeds of more than 3,500 miles per hour.

Tanks are important for battles on the ground. These vehicles carry machine guns and cannons. Militaries also use many other vehicles. Most of these vehicles have armor. Some have weapons too. But militaries do not use all vehicles for **combat**.

Russia has more tanks than any other country in the world. In 2018, Russia had more than 15,000 tanks.

DID YOU KNOW?

In 2018, the US military had more than 38,000 armored combat vehicles. That was more than any other country.

Support vehicles are not armed with weapons at all. They help soldiers complete other tasks. Some support

A US support vehicle loads a container of supplies. This vehicle can carry more than 35,000 pounds.

vehicles deliver supplies. Some carry soldiers. Others help soldiers explore land or look for enemy troops.

CHAPTER 2

HISTORY OF MILITARY VEHICLES

The United Kingdom built the first **armored** vehicle in 1900. It was a steam engine on wheels. It carried supplies. Soon after, other countries built armored cars. Some of these had guns.

LEARN MORE HERE!

In 1899, an inventor built a four-wheeled bicycle with a machine gun attached.

DID YOU KNOW?

Many ancient armies used chariots pulled by horses. Riders fought with bows and arrows.

In World War I (1914–1918), vehicles needed to be able to move over **trenches**. For this reason, militaries created tanks. Tanks used tracks instead of wheels. The tracks could grip rough ground. And a tank's gun could spin in a circle. It could fire in any direction.

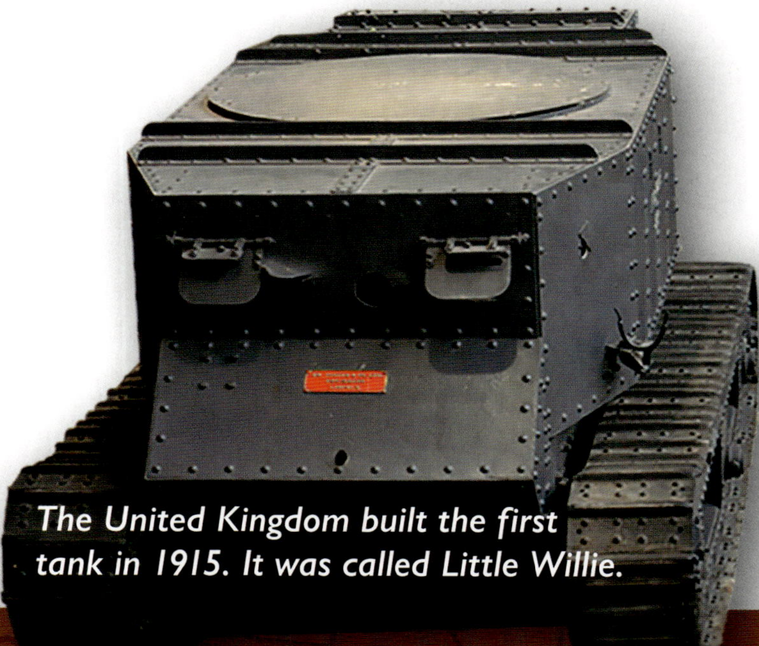

The United Kingdom built the first tank in 1915. It was called Little Willie.

A tank crosses a trench during World War I.

DID YOU KNOW?

The first tanks were slow. They could move at speeds of only 4 miles per hour.

During World War II (1939–1945), countries used faster tanks with stronger weapons. Armor improved too. At first, tank armor was made only with steel.

During World War II, Germany's tanks were more powerful than any others. These tanks were called panzers.

In the 1970s, armies added **ceramic** layers. This armor was even stronger.

The M1 Abrams was one of the first US tanks to have ceramic in its armor.

MILITARY VEHICLES

TIMELINE

1915
The UK military builds the first tank.

1800s
Bicycles are used in war in Europe.

1900
The UK military builds the first armored vehicle.

1980

The US military starts using M1 Abrams tanks.

1970s

People use ceramic layers to make stronger tank armor.

2018

Militaries begin working on a new family of vehicles.

CHAPTER 3

MILITARY VEHICLES TODAY

The US army's main tank is the M1 Abrams. Its engine is more than ten times stronger than that of an average car. The tank can hit targets as it moves. It aims with sensors and computers.

COMPLETE AN ACTIVITY HERE!

PARTS OF AN M1 ABRAMS TANK

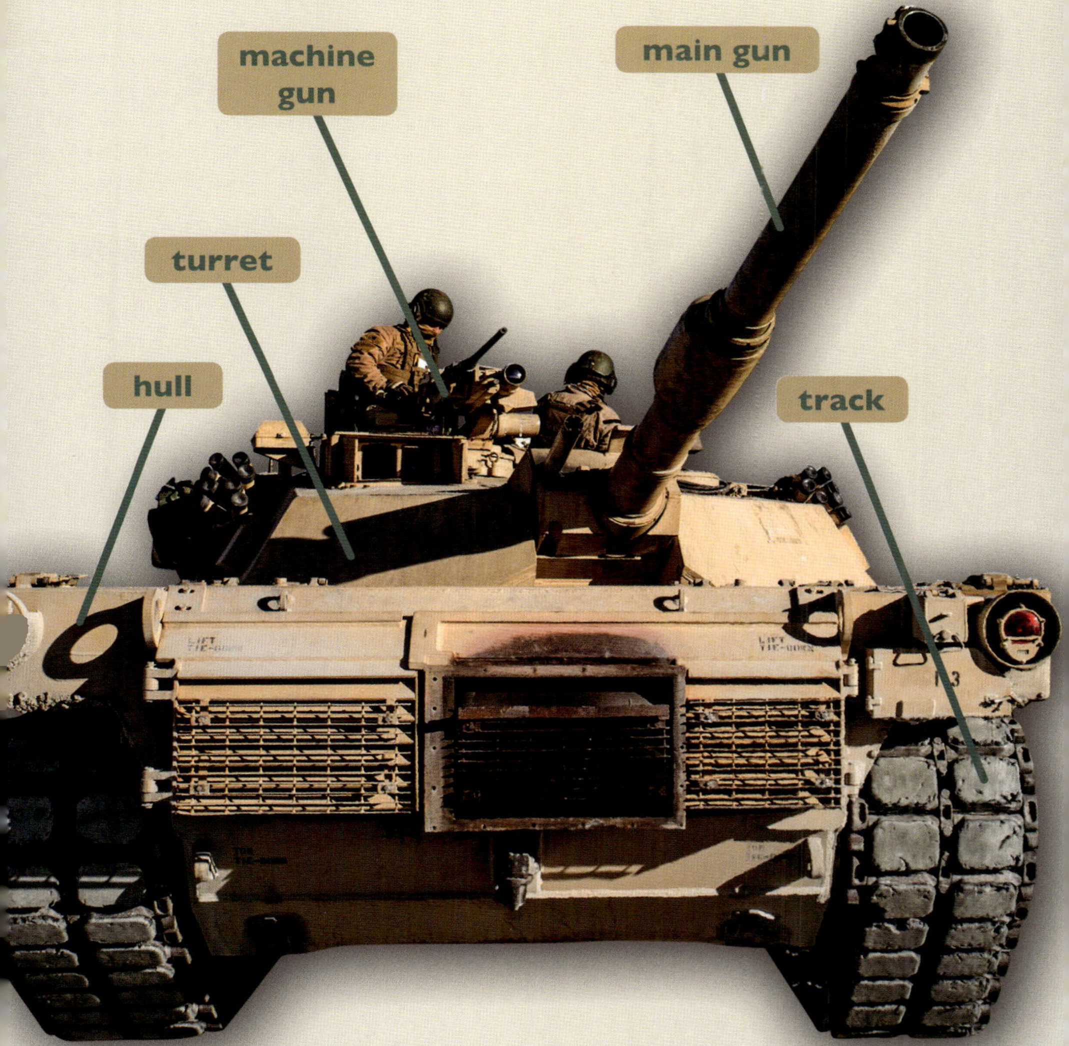

The Stryker is another type of **combat** vehicle. It is similar to a tank. But it uses wheels instead of tracks.

The US military began using Strykers in 2001.

BY LAND AND SEA

Amphibious vehicles can fight on land and in water. They often face combat near shores. These vehicles carry soldiers from ships onto land. They can travel 8 miles per hour in the water. On land, they can drive 20 to 45 miles per hour.

The Stryker can move quickly in cities and tight spaces. It can also carry soldiers.

Many support vehicles carry soldiers or supplies. Humvees can carry loads in jungles or deserts. Wreckers bring back destroyed vehicles. They use cranes

DID YOU KNOW?

The Buffalo has strong armor and a long metal arm. It clears mines and other explosives.

to pull vehicles stuck in mud or snow.

Other vehicles can even carry tanks.

Three M1 Abrams tanks weigh as much as an average house.

CHAPTER 4
THE FUTURE OF MILITARY VEHICLES

Militaries are designing new vehicles to help in even more situations. One is the Lynx KF41. Soldiers can add or remove parts of this vehicle. They can change it from a tank into a support vehicle.

LEARN MORE HERE!

Soldiers can change the Lynx KF41 into five different vehicles.

Militaries are finding many new ways to collect information. Vehicles are equipped with powerful cameras and sensors. The sensors gather lots

The US military uses computers to connect tanks on the battlefield. Soldiers in one tank can know what soldiers in other tanks know.

of information. Onboard computers help organize this information. Some computers even help vehicles complete tasks on their own.

Unmanned vehicles have no soldiers inside them. Soldiers use computers to control them from a safe distance. Unmanned vehicles can scout dangerous areas. They can go ahead of vehicles that carry soldiers.

A soldier steers an unmanned vehicle using a computer.

A vehicle drives without any soldiers at the wheel.

DID YOU KNOW?

Unmanned vehicles can fire back at an enemy if they are attacked.

MAKING CONNECTIONS

TEXT-TO-SELF

If you were a soldier, would you rather drive a combat vehicle or a support vehicle? Why?

TEXT-TO-TEXT

Some military vehicles can complete tasks on their own. Have you read other books about unmanned vehicles? What could they do?

TEXT-TO-WORLD

Support vehicles help soldiers perform many tasks. What jobs outside of the military could these vehicles help with?

GLOSSARY

amphibious – able to travel in water and on land.

armor – a hard layer that covers the outside of a vehicle and protects the people inside.

ceramic – a nonmetal material that has been heated at a high temperature.

chariot – a battle cart pulled by horses.

combat – active fighting in a war.

trench – a narrow ditch in the ground that protects soldiers.

unmanned – not controlled by someone touching the object.

INDEX

ONLINE RESOURCES

popbooksonline.com

Scan this code* and others like it while you read, or visit the website below to make this book pop!

popbooksonline.com/military-vehicles

*Scanning QR codes requires a web-enabled smart device with a QR code reader app and a camera.